Larry loves animals. He always tells me about the animals we see.

Last fall we saw a hole under a tree.
Larry said the tree was a fir tree.

Then he sniffed and said the hole was a skunk's den. After that, I left in a hurry!

On another day, Larry asked Mom for a pail. She has a number of them. He picked up her big, red wash pail. We went down to the frog pond.

First, Larry looked at each frog. "This frog is better than the others!" he said.

Larry looked the frog all over. He wrote notes on an old letter. He listed the frog's size and color.

Soon Larry let the frog go. But first he said, "Thanks, frog! You had better go home now!" The frog perched on a log.

Next year, Larry will go away to school. I hope he'll write me letters about the animals he sees.